which one doesn't belong?

which one doesn't belong?

A TEACHER'S GUIDE / CHRISTOPHER DANIELSON

STENHOUSE PUBLISHERS
PORTLAND, MAINE

Stenhouse Publishers
www.stenhouse.com

Library of Congress Cataloging-in-Publication Data

Names: Danielson, Christopher, author. | Danielson, Christopher. Which one doesn't belong? A shapes book.
Title: Which one doesn't belong? A teacher's guide / Christopher Danielson.
Description: Portland, Maine : Stenhouse Publishers, 2016. | Companion volume to: Which one doesn't belong? A shapes book. | Includes bibliographical references.
Identifiers: LCCN 2016018737 (print) | LCCN 2016031466 (ebook) | ISBN 9781625310811 (pbk. : alk. paper) | ISBN 9781625311306 (ebook)
Subjects: LCSH: Shapes--Study and teaching (Elementary) | Geometry--Study and teaching (Elementary)
Classification: LCC QA461 .D26 2016 (print) | LCC QA461 (ebook) | DDC 516/.15--dc23
LC record available at https://lccn.loc.gov/2016018737

Book design by Tom Morgan (www.bluedes.com)

Manufactured in the United States of America

 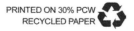

For Wendy: my closest and strongest role model for intellectual and emotional strength, and for critical thought

CONTENTS

Acknowledgments

I am deeply indebted to Megan Franke and Terry Wyberg—two amazing teachers without whom I never would have encountered the *Which One Doesn't Belong?* concept.

I am grateful for Toby Gordon, Dan Tobin, and Tracy Zager—this Stenhouse team has been crucial to bringing this project to fruition.

I want to thank the teachers and students who have welcomed me into their classrooms as I wrote and revised this book, and the teachers and students who have welcomed my ideas and questions into their classrooms via the Internet.

Which One Doesn't Belong? on the Web

You can find a variety of resources for *Which One Doesn't Belong?* on the website www.stenhouse.com/wodb. Play with the zoomable heart from Chapter 4, find additional *Which One Doesn't Belong?* sets for use in class, and use the sign-in information below to download high-resolution images of all the pages from the student book. You can also join the conversation on Twitter using the hashtag #wodb.

SIGN-IN INFORMATION
To access the book's digital content on www.stenhouse.com/wodb, please enter the code DELIGHT.

CHAPTER 1
Why Another Shapes Book?

I am a math teacher and a father. Many shapes books have come through my home. Many have been fine; some have been mathematically correct (though certainly not all of them!). None has offered my children and me an opportunity to talk, to argue, or to wonder.

For a number of years I have longed for a better shapes book. I have wanted a shapes book that gives space for noticing relationships, for asking questions, and for thinking together. I designed *Which One Doesn't Belong?* to be such a book, and now I offer its collections of geometric objects to you and your students. Together, they are an honest and complete invitation to mathematical conversation.

As with all ideas, this one owes a debt to the inspirational work of others. Many readers will be familiar with *Sesame Street's* "One of these things is not like the others" segments. In those segments, there are four things—three are alike and the fourth is different. In one such segment, Grover thinks hard about four circles, three of which are large while one is small. An important property of these *Sesame Street* segments is that they have one unique answer. One thing really is different from the others in a predetermined way.

Math educators have adapted this activity to build engaging classroom interactions. Megan Franke at the University of California, Los Angeles, for example, has designed sets of four numbers in such a way that any one of the four can be seen as different from the other three. This small adjustment takes a simple activity about noticing sameness and difference and transforms it into a challenging task that supports rich conversations requiring precise language use and sophisticated argumentation.

To better understand the importance of this transformation, consider the difference between the two examples in Figure 1.1. In the example on the left, one thing is unambiguously unlike the others. In the example on the right, there are many attributes that vary; this leads to many ways of identifying which one doesn't belong. In one example, you notice that all are the same shape but one is a different size. In the other example, you may pay attention to shape and size, but also to orientation, shading, and composition. You are done with the first example fairly quickly. The second example offers a moment of surprise (Wait! There's more than one right answer?) that draws you in more deeply and invites you to linger for a while and think.

FIGURE 1.1
Which One Is Not Like the Others? at left; *Which One Doesn't Belong?* at right

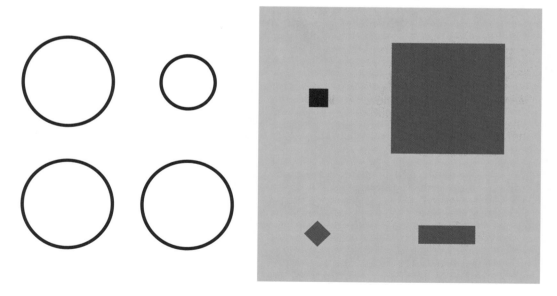

Here is a sampling of what a typical classroom of first graders has to say about the example on the right:

The shape in the upper left doesn't belong because it is blue, while the others are red.

The shape in the upper right doesn't belong because it is bigger than the others.

The shape in the lower left doesn't belong because it is standing on its point.

The shape in the lower right doesn't belong because it can't be a square.

In this book, the question "Which one doesn't belong?" is ambiguous; there isn't just one answer. In fact, any one of the shapes on any one of the pages can be the one that doesn't belong. All choices are correct, which shifts the focus to justification. *Which One Doesn't Belong?* isn't about guessing the right answer; it's about expressing mathematical relationships precisely in order to communicate with others.

Commonly in math class, student responses are compared to a standard answer key—the measure of what's right is what's in the back of the book, or what the teacher has in mind. In a conversation about a well-designed *Which one doesn't belong?* task, the measure of what's right is what's true. While I do provide you with possible answers, their purpose is different from an answer key. I provide possible answers to help you anticipate the wide variety of properties students will notice, and maybe to help out when you and your students are nearing frustration with one or two shapes on a page. But please note that being listed among the possible answers is not the measure of what's right. Instead, what's right is determined by what's true. Does everyone agree that the difference you're pointing to exists? Then you're right! You may not be using the right words for that difference, but there's time to develop vocabulary later. You may not be noticing something from the standards at your grade level, but that's okay, too. If what you notice is true and if you can communicate it to others, then it's right and it will help others to expand their understanding of these geometric figures. Of course, wrong (i.e., untrue) answers are welcome contributions as well. These provide opportunities for clarification, argumentation, and perhaps some impromptu measurement (*Is that really a right angle? Are those two sides parallel? How could we check? How can we know for sure?*).

Which One Doesn't Belong? is a book about shapes. More generally, it's a book about mathematics. When children look for sameness and difference; when they work hard to put their ideas into words; when they evaluate whether somebody else's justification makes sense; when they wonder, *What if we rotated that one?*—in all of these cases, children engage in real mathematical thinking. They build mathematical knowledge—knowledge they can own and be proud of. They develop new questions. They argue. They wonder.

In this teacher's guide, I describe some important things we know about children's learning of geometry, relate this research to the design and implementation of the tasks in *Which One Doesn't Belong?* (including challenging examples from classrooms at several different grade levels), and look beyond these tasks to extension activities and ways to continue the conversation beyond geometry.

Chapter 2 focuses on research and the role of geometry in the mathematics curriculum. In Chapter 3, attention turns to the role of this book in classrooms. How do I use this book? Where does it fit in with what I am already doing? Why would I use this book? When? How? Chapter 3 considers these important questions of implementation. Chapter 4 works on the relationship between children's mathematical thinking and formal mathematics. For this reason, Chapter 4 is the most important chapter in the book. In Chapter 5, I get specific and offer possible answers for each page in the book. This final chapter offers the reader a view into the beautiful and creative minds of diverse children. All in all, this teacher's guide includes not just possible answers but some sample conversations. The goal is to show how mathematical discussions can move forward, and how ideas can build on each other as children consider new ways to look at shapes. Throughout, I'll share my journey as a teacher with these ideas as well as wisdom from teachers who have reported their experiences to me.

CHAPTER 2
How Children Become Geometers

Geometry has occupied a strange place in US elementary and secondary curriculum for quite some time. Typically in elementary school, students have learned tremendous amounts of vocabulary. *Right, acute, obtuse, scalene, equilateral*, and *isosceles* are only some of the terms that apply to triangles, which themselves constitute only a small corner of the world of geometric figures. After the vocabulary work of elementary school, most students study relatively little geometry until high school, when they take a yearlong geometry course focused on proof.

In this chapter, I provide a richer picture of what geometry learning can look like, drawn from research and the Common Core State Standards.

THE VAN HIELE MODEL

The van Hiele model is foundational research for understanding the growth of children's geometry knowledge. Pierre van Hiele and Dina van Hiele-Geldof developed this model as they sought to understand and describe how children's geometric thinking develops over time (van Hiele 1985). Theirs is a big-picture model. It isn't useful for making fine-grained assessments of student learning, but is instead very useful for describing categories of thinking you may see in any elementary or secondary classroom, and for planning future instruction.

The van Hiele model has five levels, numbered 0—4. Each level describes a category of student thinking about geometry, and the levels are hierarchical. This means that students thinking at a higher level of the model must have had experiences thinking at the lower levels too. In this sense, lower levels are contained within the higher ones.

Level 0: Visualization. At level 0, children pay attention to what a shape looks like. *A rectangle looks like a door* is an example of level 0 thinking.

Level 1: Analysis. At level 1, children notice properties of shapes and begin to develop vocabulary for these properties. *A rectangle has four sides and all right angles* is an example of level 1 thinking.

Level 2: Informal deduction. At level 2, children begin to build arguments about relationships between properties of classes of shapes. *Like a rectangle, a square also has four sides and all right angles, so a square is a special kind of rectangle* is an example of level 2 thinking.

Level 3: Formal deduction. At level 3, students support their claims systematically with chains of logical reasoning. Mathematicians call these chains of reasoning proofs. The paragraph in Figure 2.1 is an example of a proof of the claim that the sum of the measures of the interior angles of any triangle is 180°.

Level 4: Rigor. At level 4, students look beyond the constraints of the familiar and consider alternative geometries. *How is geometry different on the spherical surface of the Earth than it is on an infinite flat plane?* and *How is geometry different if—like a taxicab navigating a city—there are no diagonal lines but only right angles?* are the kinds of questions people investigate at van Hiele level 4.

Typically, American curriculum has lingered on levels 0 and 1 in the elementary grades, where students sort shapes based on number of sides and learn many vocabulary terms. Little instructional time is spent at level 2 before students are plunged headlong into the level 3 work of high school geometry, where many students struggle to understand the purpose and structure of mathematical proofs. Van Hiele level 4 has been the preserve of undergraduate math majors and beyond.

In recent years, some of the stronger elementary and middle school curricula—in particular those developed following the *Standards for School Mathematics* (NCTM

There are two facts about angles we need to know in order to prove this. First, if you stand in one place and spin around exactly once, you will have turned 360°. Second, two angles that share a side and combine to form a straight line together measure 180°. Now, imagine that you are going to walk around the triangle pictured in Figure 2.1, starting at vertex A and facing vertex B. You walk to B, where you turn left to face C. You walk to C and turn to face A. You walk to A and turn one last time to face B again. The pair of angles shown at each vertex (e.g., A1 and A2) share a side and form a straight line, so their sum is 180° in each case.

There are three of these pairs, so they total 540°. At each vertex, you turned through the angles marked A1, B1, and C1, respectively. You started and finished your walk facing vertex B, so in the process of walking around the triangle, you rotated 360°. When you subtract the measures of A1, B1, and C1 from the total, you are left with the sum of the measures of A2, B2, and C2, which is 180°. But there is nothing special about this triangle. Everything I have described about this triangle would be true of any triangle. Therefore, the sum of the interior angles of any triangle is 180°.

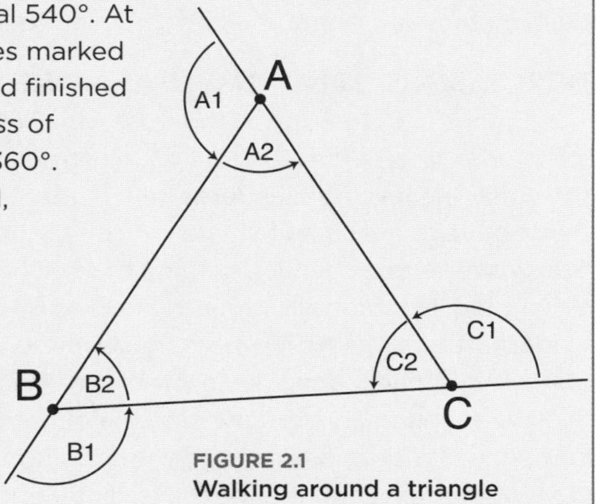

FIGURE 2.1
Walking around a triangle

1989)—have incorporated a lot of meaningful work at level 2 of the van Hiele model. Nonetheless, the experience of many American students has been that high school geometry is an abrupt change rather than a natural evolution of their prior geometry work.

The van Hiele model provides a structure for understanding this all-too-common experience. In order to build formal and logical arguments (at level 3), you need to have practice making informal arguments (level 2). When students don't have that practice prior to high school geometry, proof writing becomes more of an exercise in trying to guess what the teacher wants, rather than exploring the forms and constraints of logic. The van Hiele model is based on the idea that you need experience at each level before you can move to the next, and that instruction is a necessary ingredient at each level.

The van Hieles haven't argued that it is *difficult* to go from level 1 thinking directly to high school geometry; they have argued that it is *impossible*. If students don't have experience and instruction building informal geometry arguments, they will not learn to write proofs. They may learn to imitate the form of a two-column proof, but they will not build mathematical arguments as they do so—they won't understand what they are doing.

DESCRIBING STUDENT LEARNING WITH THE VAN HIELE MODEL

The Common Core State Standards, and those of many non–Common Core states, are structured to provide students with instructional experiences that progress through the van Hiele levels. In kindergarten, students identify and describe shapes. They describe relationships of shapes to each other and in space. *The bigger circle is above the smaller one* is the sort of thing many kindergartners need to practice noticing and describing. Much of this is van Hiele level 0 work. Note that this does not mean that kindergartners' geometry learning is limited to naming shapes and describing which is above the other. Indeed, an important premise of *Which One Doesn't Belong?* is that even very young children are capable of noticing and (to a lesser extent) describing complex relationships among geometric figures. Work at levels 0 and 1 is important for developing the language to discuss and critique these sophisticated ideas.

Across the primary grades, children extend this work by looking for similarities and differences among groups of shapes. They consider the meaning of such words as *right, square*, and *angle*. Often this learning requires revision, and then re-revision, of previously understood ideas.

Writing a definition that is both true and complete is challenging level 1 work. *A rectangle is a four-sided shape* is true but not complete. Even a child who can correctly pick the rectangles out of a set of quadrilaterals may still struggle to state precisely what makes those shapes rectangles. Similarly, a student who has developed a complete definition of a rectangle (for example, a four-sided polygon with all right angles) may not recognize that this definition allows squares. Students often rely on visualized examples rather than logical definitions, and the image of a square as unique may override the definition the child knows. This is normal and important level 1 work.

Even while students continue to wrestle with these level 1 ideas, they also begin to work on relationships among the properties they work with. An extended discussion of the shapes in Figure 2.2 could result in a student noticing that whenever triangles have all their sides the same length, the angles are the same too; but that's not true for quadrilaterals. This is a level 2 observation because the student describes a relationship between properties for classes of shapes, not just the particular shapes in front of them. You could rephrase the preceding claim this way: all equilateral triangles (those with all side lengths the same) are also equiangular (they have all angles the same size). But not all equilateral quadrilaterals are equiangular. The relationship between the properties *equilateral* and *equiangular* is not the same for these two classes of shapes. At van

Hiele level 2, students look beyond the particulars of the shapes they are looking at, and they look across different categories of shapes to make and defend claims about geometric relationships.

WHICH ONE DOESN'T BELONG? TO THE RESCUE!

A well-executed *Which One Doesn't Belong?* conversation is an excellent way to strengthen the van Hiele levels 1 and 2 instruction in any classroom. As an example, consider the spirals in Figure 2.3.

FIGURE 2.2
All of these shapes are equilateral. They are not all equiangular.

FIGURE 2.3
Spirals

On one of my first classroom visits to share this book, a first grader made an interesting observation about these spirals. "None of these belong," she said, "because on the other pages the shapes were either colored in, or they could be colored in. You can't color these in even if you tried."

This girl was building an informal argument about shapes that are closed and those that are not. A polygon is defined as a simple, closed plane figure with straight sides (see Figure 2.4).

The spirals prompted this first-grade girl to offer a truly useful definition of the term *closed*. Her implicit definition is phrased in perfect language for young children to understand: a closed figure is one that can be colored in (or perhaps that is already colored in). This first grader didn't know that closed is the word mathematicians use to describe this property, but that doesn't matter. The *Which One Doesn't Belong?* conversation gave her an opportunity to practice making claims about collections of geometric figures, which is van Hiele level 2 work for her. At the same time she provided level 2 instruction for her peers. Every first grader in that classroom was evaluating her claim by considering whether it is possible to color in spirals.

Argumentation at level 2 makes up some of the richest, most interesting work of elementary and middle school geometry. While this first-grade class was just getting started with this kind of argumentation, upper grades need to practice flexing these particular mathematical muscles. When I told the story of the spirals to a fifth-grade classroom, one student argued that the claim was false. She said, "Those spirals are very

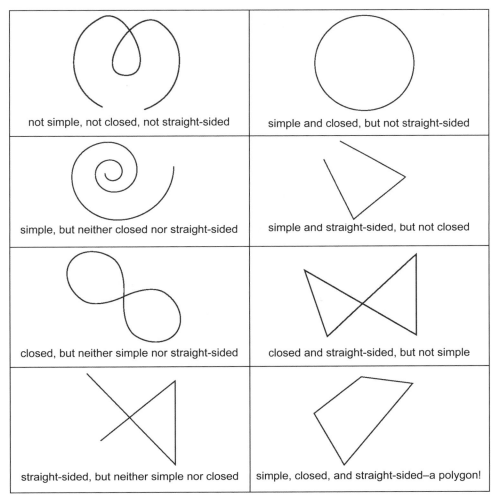

not simple, not closed, not straight-sided	simple and closed, but not straight-sided
simple, but neither closed nor straight-sided	simple and straight-sided, but not closed
closed, but neither simple nor straight-sided	closed and straight-sided, but not simple
straight-sided, but neither simple nor closed	simple, closed, and straight-sided–a polygon!

FIGURE 2.4
Many shapes with many combinations of important properties, but only one polygon

long, very skinny rectangles that are rolled up—if you unrolled them, they'd look like long, thin rectangles. And they'd be colored in." This is a delightful counterargument! This sort of back-and-forth can happen when students know that ideas are for considering, and for accepting, rejecting, and modifying. Students can listen to each other's geometry claims. They can evaluate these claims, and they can argue with evidence when they disagree. This ongoing level 2 work is an important part of advancing in geometric sophistication, and it's an important part of becoming a better mathematician.

As you use this book with students, and as you and your students begin to craft your own *Which One Doesn't Belong?* sets, keep an eye out and an ear open for the subtle differences between considering the individual shapes we're looking at and using these shapes as examples of whole classes of shapes. These shifts are characteristic of progressing through van Hiele levels, and they can signal deep learning. Yet these are subtle shifts sometimes signaled only by a word or alternative phrasing. Assessment in teaching—as distinct from grading—is about characterizing student knowledge. It requires listening carefully to students in order to understand their perspective, to learn what they know. As you listen to your students describing which one of several shapes doesn't belong and explaining why, the van Hiele model provides a useful framework for understanding what your students know, and for thinking about how to help them further their learning.

CHAPTER 3

How to Use *Which One Doesn't Belong?* in Your Classroom

I am in a classroom, in front of the interactive whiteboard with 23 first graders sitting on the carpet at my feet. I have invested many hours in developing this book and now is a moment of truth. My own children and I have had a ball at home, looking at the shapes and talking about them. I have revised the individual sets of shapes, thrown some away, and added new ones. I know that *I* love this book. But will it work in a classroom full of children whom I have never met?

> We are going to look at a book together. It's a book of questions, but there is only one question that's the same on every page: "Which One Doesn't Belong?" On each page, you'll see four shapes—one in the upper left, one in the upper right, one in the lower left, and one in the lower right.

I point to each of these spaces as I talk. I know that my second-grade daughter hasn't nailed down these spatial relations yet, so I don't expect these first graders will have either. But I also know that repeated exposure is how children learn language. So while I don't expect them to talk about the "upper left" shape, I will use that language and also provide the gestures to support it.

> The way this will work is I'll show you those four shapes, and I'll ask you to think to yourself which one seems different from the others, and why. I'll ask you not

to say anything at first so that everyone has a chance to think. But then I'll call on someone to talk to us about which shape they chose. Our goal will be to understand how that person is thinking. Then I'll call on someone else—hopefully someone who chose a different shape, and probably for a different reason. The great thing about these collections of shapes is that I designed them so that you can pick any shape on the page and find a reason for it not to belong. Every shape is like the others in some ways, and different from them in some other ways. So this isn't a game where you are trying to guess the right answer, or to figure out what I was thinking when I drew these shapes. Instead, this is a game where you try to see new things in these shapes—partly through your own ideas, and partly by listening to your classmates.

On the first page, we'll make sure we get to each of the four shapes. Then we'll move to another page that will have new shapes with new relationships for us to think about. We'll do four or five pages together in the next twenty minutes, and I hope I'll learn something new from you about these shapes. I bet I will. So let's go.

Having set up the things that are important to me—quiet thinking time, multiple correct answers, a focus on explanation, an invitation to see things in new ways, and a brief statement that I hope to be a learner too—I show them the first page of the book (see Figure 3.1).

The students talk to me about triangles, diamonds, rhombuses, squares, angles, orientation, composition, and probably several other things that I cannot specifically recall five minutes after the lesson because I am just overwhelmed by the children's focus, enthusiasm, and ideas. I am delighted that several students want to share a second reason for a shape that we have already discussed, saying things such as, "I chose that shape too, but for a different reason."

We do a second page. This time, I have them turn and talk to their neighbors before sharing in the large group. I want everyone to have practice voicing their ideas. I ask the class to help make sure a variety of different students get a turn to talk in the large

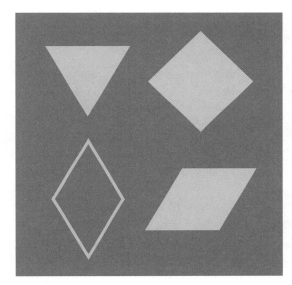

FIGURE 3.1
The beginning of *Which One Doesn't Belong?*

group. As a former middle school teacher, I inwardly marvel at the differences between managing a group of middle schoolers and a group of first graders (to say nothing of the differences from the college students I have been working with for the past ten years!).

I ask the students whether the third page we do should be like the ones we have done, or whether we should do a challenge page. In a scene that ends up playing out in dozens of classroom over the next few months, the children loudly cheer "Challenge page!" So we skip ahead to a page with a bunch of shapes that aren't usually in a geometry textbook (see Figure 3.2). They *ooh* and *aah* at this page, but more importantly they have really interesting things to say. We talk about corners and vertices, whether curves count as sides, and what the shapes would look like if you cut them apart.

At each turn of the page I summarize what we have discussed, and I encourage them to reuse these ideas and to develop new ones as we consider new collections of shapes. After a fourth page, my time is up. I have scheduled another classroom, but I don't get the sense these students are done with these ideas. While it's a good idea to move on before boredom sets in, the variety of shapes and the challenging relationships have held their attention and interest. Nearly everyone has spoken in the large group, and they have certainly all shared ideas with their neighbors.

I leave for the next classroom with questions about how to improve the experience for all students, including these:

FIGURE 3.2
Children enjoy a challenge.

• Whose ideas get picked up and used by other students?
• How do I make sure we hear from everybody?
• How do I decide which ideas to introduce if students don't bring them up themselves?

I know that these are things experienced elementary teachers can help me with as I visit more classrooms in the coming weeks and months. My big question, "Will children who are not my own be interested in these collections of shapes?," has been answered. These shapes have engaged their minds and brought out complex ideas that are much more interesting to all of us than if we had spent all twenty minutes counting sides on polygons. I have work left to do, but I leave this classroom knowing that I have built a resource that can make a meaningful contribution to geometry instruction in real classrooms.

THE PEDAGOGY OF *WHICH ONE DOESN'T BELONG?*

An important part of the message of *Which One Doesn't Belong?* is that quality math instruction need not be so different from other kinds of teaching. Many teachers who feel well equipped to bring literature, history, or science alive in their classrooms don't see the same kinds of opportunities in mathematics. If these other subjects are exciting because learners can see themselves in the stories and studies, then helping students see themselves in mathematics can be similarly effective and engaging. In helping you to think about the roles that *Which One Doesn't Belong?* can play in your classroom, I'll begin with an analogy about a terrific science project that came into my home.

My daughter, Tabitha, is eight years old and in third grade. She was working on an animal project for school recently in which she needed to choose an animal, research it, write about it, and build a model of its habitat. Tabitha chose flamingos for her project— the Andean flamingo, to be precise. One evening, she showed me a delightful website she was using for her project. The website is called ARKive, and consists largely of short videos of animals in the wild, organized by animal. We spent a good twenty minutes looking at the Andean flamingo page together. We had a great time, and by the end I felt as though I knew some fascinating things about a creature I wouldn't have expected to find interesting.

Reflecting on the experience, I wondered why I had found those twenty minutes so much more captivating than many other experiences I have had on the web. I wondered why I had questions about Andean flamingos that I had never before thought to ask. I wondered why I felt I had learned something in a way that I often don't when I go to a large-screen, highly produced film at a science museum.

The videos on ARKive.org are not narrated or captioned. They are short clips, lasting from thirty to ninety seconds, of various behaviors of the animal in question in its native habitat. Tabitha's and my experience was mediated by cameras, and by editing and clip selection. But this mediation is all in the service of bringing together interesting behavior for us to observe. There is no narrator telling us what to notice or what questions to ask. No one told us what to think about what we saw. The absence of narration and explanation allowed us to play with ideas and associations such as the airplane-like landing habits of these very large birds.

If you have ever watched airplanes land on a runway, you can probably see in your mind's eye the image of these large machines descending slowly, but moving forward at a high speed, wheels outstretched to the ground. Then the wheels hit the ground and the plane slams on the brakes, coming to a stop while covering some distance. That's what Andean flamingos do. They approach their briny shallow-water runway low and fast, reaching one foot at the end of a long leg out tentatively, and transitioning to a run once that foot catches the ground. They slow their run and come to a rest after having covered some significant distance on the ground. Tabitha and I laughed with delight at this video, and I made her watch three or four times while I contrasted it with how ducks land (semi-graceful splashdown in the water). Pretty soon I was doing an elaborate impression of a heron landing on our local lake, flapping my wing-arms wildly as I came to a near standstill in the air and settling down vertically into the knee-deep water. For her part, Tabitha snapped her head back and forth in imitation of what we took to be the male Andean flamingos' group mating dance. A video of a flamingo chick had us wondering about the shape of the baby's beak in comparison to that of the adult, and the care with which the adult must manipulate its enormous beak while tending to such a tiny chick.

I spent some time the next day wondering about how other birds land. How do seagulls end their flight? This kind of lingering curiosity, stemming from newly noticed properties, is precisely what I hope teachers will be able to provoke through mathematics teaching in general, and through *Which One Doesn't Belong?* in particular.

In the student book, I have curated geometric properties in a similar manner to ARKive's curation of animal behaviors. A lot of work has gone into the selection and organization of the images in these pages so they stimulate conversations about important geometric ideas, but I have not told students what to think. The book contains no definitions and no vocabulary.

In the conversations I have had in classrooms, I have seen students moved to the same kinds of explanations I provided Tabitha when I wanted to talk about the different landing styles of large aquatic birds. Children draw triangles in the air, tip their heads at just the right angle, gesture, and make impromptu noises to punctuate the actions they imagine performing on the shapes we discuss. They ask questions without expecting an immediate answer. They imagine new shapes that aren't in front of us and bring these examples to our attention so that we can better understand their viewpoint and their inquiry.

This chapter gives you some of the basic tools and perspectives that are important to making *Which One Doesn't Belong?* a rich and impactful resource in your own classroom. Whether you are a new teacher or a longtime veteran, experienced with open and rich mathematical tasks or preparing to take a few first tentative steps with mathematical inquiry in your classroom, this chapter offers relevant experience and wisdom from the communities of teachers that have supported and nurtured my own inquiry in mathematics learning over the last twenty years.

General Guidelines

While teachers, students, and classrooms vary in needs, tone, and culture, there are a few guidelines that are important to conducting successful *Which One Doesn't Belong?* conversations. These guidelines apply whether you are a middle school teacher with a group of 32 seventh graders, a kindergarten teacher with 24 five-year-olds, or a parent sitting down on the couch for an evening book with your four-year-old. These guidelines are important considerations for establishing and maintaining a safe space for sharing and learning from each others' ideas.

WAIT TIME

Everybody needs a chance to think, and *Which One Doesn't Belong?* has the potential to trigger especially deep thoughts that students need time to work out in their own minds before speaking aloud. To allow for this, you can ask students to keep their ideas to themselves for a specified time while everyone has quiet thinking time. Discourage them from raising their hands until the allotted time has passed.

LOWER THE RISK

In a *Which One Doesn't Belong?* conversation, you will encourage students to take intellectual risks by talking about differences they notice but which they may not have a name for. This means that all claims need to be taken seriously and judged on their merits. It also means that the teacher needs to let go of a little bit of power for judging correctness; this is a job for the community of learners to do with grace and respect. It does not, however, mean accepting utter nonsense as truth. The examples later in this chapter are intended to help clarify techniques for navigating these sometimes-murky waters.

LEAVE QUESTIONS OPEN

It is sometimes important to leave unanswered the questions a class is considering.

Maybe your class has a good reason for each of three shapes on a page not to belong, but no one has a reason for the fourth one. It's OK to close the conversation and move on without a solution being voiced. Or maybe there is disagreement about the meaning of a term. If you can clarify the question with the class and post a visual reminder for students to ponder, you will frequently get new, thoughtful responses from students over the course of the next hour, day, or week.

DIG DEEP

This advice may seem to be somewhat the opposite of the preceding *leave questions open* exhortation, and in truth a balance is important. It is especially important early in a class's experiences with *Which One Doesn't Belong?* to encourage students to dig deep—to go beyond their first impressions and try to find (1) reasons for more than one shape not to belong and (2) a second reason for an individual shape not to belong. The early pages of the book are structured in ways that invite this more readily than later ones, where the similarities and differences tend to be more subtle and complex. For this reason, it's a good idea to begin with the first set in the book—no matter what age group you're working with. Kindergartners and middle schoolers alike can see and talk about the differences there without feeling intimidated, overwhelmed, or (in the case of the older students) condescended to. There is enough richness so that various age groups can dig deep. Depending on your students and their experiences, feel free to pick and choose the sets you explore next. I'll discuss guidelines for making choices later in this chapter.

PURPOSES

I once attended a wonderful workshop by Malke Rosenfeld, a talented colleague who works with children in math, motion, and dance in a program she has developed called *Math in Your Feet*. I saw rich connections to important mathematical principles as I worked with a partner to develop a short percussive dance routine within Malke's constraints. At the end of the session, a facilitator asked participants to answer the question, "What did you learn today?"

I struggled at first because this question—to me—seemed to require a declarative statement. "I learned that even numbers are divisible by two" or "I learned how to factor quadratic equations"; something of that nature seemed called for. Yet the workshop hadn't been about a particular skill or set of facts. If I had to state in a declarative sentence what I had learned that morning, I worried that maybe I hadn't learned anything.

This experience was transformative for me as a learner and as a teacher. I came away from it with a deeper understanding of what I value in learning. Of course being able to state new facts is an aspect of learning, but much more important to me is being able to ask new questions. The new facts I could state after the morning's work were mostly trivial, and nearly all were particular to *Math in Your Feet*. But the questions I began to ask as a result of the morning's work sparked new questions and entire new lines of inquiry in my teaching and learning. These questions have had an impact far greater than any small set of testable facts could have had.

Some examples of new questions that arose for me as a result of Malke's workshop are: What is the relationship between variable and attribute? Is decomposing things by their attributes—and then paying attention to one of these attributes at a time—a practice unique to mathematics? Is it a characteristic of a novice that he or she is unable to distinguish noise from pattern?

Once I characterized learning as having new questions to ask, I began to think about my students' learning differently, and I began to think about my assessment practices differently. Where in my quizzes, tests, and other assignments was I checking whether my students had new questions to ask? What does it mean to teach mathematics in a way that sparks new questions for students?

Which One Doesn't Belong? is a partial answer to this last question. Each of these sets of four objects provides a space to ask questions, speculate, and wonder about properties, relationships, and ways of seeing. Some of the questions that *Which One Doesn't Belong?* conversations have generated from students include

"What counts as a shape?"

"What is the difference between a corner and a vertex?"

"How can we measure the length of a spiral?"

"Is a square the lines going around the outside or the colored stuff inside those lines?"

"How do we know which of these properties matters?"

These are important, profound questions that cut to the heart of mathematical inquiry. As students discuss and otherwise explore the answers, they make claims that develop into knowledge. But instead of trying to commit someone else's knowledge to memory, students learn the story of their own ideas. "I used to think only things with straight sides counted as shapes, but now I know that there's a special word for that—*polygon*—and that lots of other shapes have interesting properties too" is the sort of learning that stays with students over the long run.

While I am sensitive to the data-driven environment that constrains teachers' time and task selection, I have seen repeatedly that well-chosen, open-ended tasks can profoundly change the ways that children view their intellectual and physical worlds, just as Malke's workshop forever altered my own views on teaching and learning.

Yet the kind of teaching that builds on students' ideas and questions isn't lacking direction or purpose. Indeed, I advocate being very clear about the purpose of a *Which One Doesn't Belong?* conversation. Next I outline a few important purposes, including samples of the ways teachers can talk with students directly about each purpose and teacher moves that will help to achieve these ends.

Defining Terms

One of the distinguishing features of mathematics as a discipline is a reliance on clearly defined terms. In contrast to literature—where ambiguity and subtext are often desirable characteristics of the work—mathematics requires everyone to understand the same thing by a term. Importantly, this shared and precise understanding is the end product of mathematical activity, while traditional textbook instruction treats it as the beginning point.

There are many great examples in the history of mathematics that illustrate this point. Sir Isaac Newton developed his version of calculus by relying on *fluxion*, a term that had a vague meaning but was refined over many years and by many people into *limit* and *derivative*. Imre Lakatos (1976), in his book *Proofs and Refutations*, has documented the development of the term *polyhedron* as a process of mathematicians arguing over properties of these three-dimensional figures. In a key part of this process, various mathematicians presented counterexamples (referred to in the text as "monsters"). Each monster is a figure that meets the then-current definition but does not have the property in question. At each stage, the mathematicians refined their definition of polyhedron. At the outset, polyhedron is a poorly defined term that allows for disagreement about which objects are polyhedra and which are not. Through careful analysis of examples, the definition becomes more precise. Precision of language is a product of doing mathematics, not the origin.

So it can be in classrooms. Imagine you are a primary teacher beginning a geometry unit with the study of triangles. You could very early on ask students, "What is a triangle?" English-speaking schoolchildren of nearly any age will have something to say about this—with possibly wide-ranging levels of sophistication and evidence of van Hiele levels. Second-language students may require more support in recognizing the term and in expressing their ideas. In most K—2 classrooms, a quick conversation will likely build consensus around the idea that a triangle is a shape with three sides. Like Newton's fluxions and the polyhedron that Lakatos describes, this definition of triangle is not precise enough to sort out the real triangles from the impostors.

Figure 3.3 shows a *Which One Doesn't Belong?* page that is designed to elicit arguments that will help students refine their definition of a triangle. After the initial discussion in which students share their current understanding of triangles, you could say something such as, "Right now, we're thinking that a triangle is a shape with three sides. We will come back to that in a little while. First, I want us to look at a set of shapes together. When you see this set of shapes, I want you to think to yourself, *Which one seems different from the others?* and *How would you tell someone else about the difference you see?* Don't say anything out loud at first. Think it to yourself. Then, in a minute, you'll tell someone else which one you picked and why, and you'll listen to what they have to say about the shape they picked."

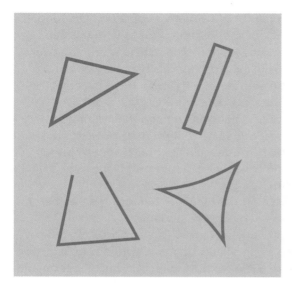

FIGURE 3.3
A page of "triangles"

At some point in the ensuing conversation, a typical primary-aged classroom will disagree about whether the shape in the lower right is a triangle. It isn't a triangle, but the importance of straight sides isn't obvious to most young children. If your students disagree about this shape, it may be an opportunity to leave a question open for a little while as you give them some additional experiences with triangles. If you have a classroom full of students who all believe the lower right shape to be a triangle, you can plant a seed of doubt yourself by saying something such as this: "We all agreed earlier on that one reason for this shape not to belong on this page is that it has curved sides. Now it seems we are all in agreement that it is a triangle. So I wonder whether it's OK for a triangle to have curved sides? Let's think about that over the next couple of days, and let's look for other examples of triangles with and without curved sides."

In Chapter 4, you'll see how students have generated definitions for terms such as *convex, concave, vertex, diamond*, and *square*.

Introducing New Ideas

Which One Doesn't Belong? is a useful structure for getting a group of students unstuck. I was visiting a kindergarten classroom over the course of a year, once or twice a week during math time. Early in the school year, the kindergartners were studying patterns. I was worried that the instructional materials might conflict with their informal understanding of pattern by focusing exclusively on one kind of patterning—repetition of some elementary combination of letters, colors, or shapes. For instance, an AB pattern repeats two elements: black-white-black-white and so on. An AAB pattern might be orange-orange-blue-orange-orange-blue and so on.

I know that there are lots of patterns that can't be expressed as strict repetitions of basic elements. So I borrowed some snap cubes and went home to build a *Which One Doesn't Belong?* prompt (Figure 3.4) to introduce some new patterning ideas to these students.

FIGURE 3.4
Which pattern doesn't belong?

Students picked up the color in the lower right, the orientation in the lower left, and the two red cubes in each part of the repeating pattern in the upper right. But they struggled to say how the one in the upper left was different. They could see it, but they couldn't say it. They kept reading the pattern red-white-red-red-white-white and so on. After three or four minutes of different students trying to describe how this pattern is different, one girl said "There's one of each, then there's two of each . . . that's changing, but it's the same in the others." A couple of weeks later, during free-play time, another kindergartner was very carefully building this pattern block pattern: hexagon-square-hexagon-hexagon-square-square, and so on. That one conversation succeeded in helping her to think in new ways about patterns in math class.

I give advice for building your own *Which One Doesn't Belong?* sets at the end of this chapter, and I strongly recommend playing around at this—you'll learn some math, and you'll learn a lot about how students think. But the sets in this book are designed to give you chances to introduce important geometry ideas without having to tell students about them in advance. Students naturally talk about straight and curved sides on page 21; they naturally count sides and attend to the orientation of shapes on page 5; they nearly always talk about symmetry on page 19.

I chose the sets of shapes in *Which One Doesn't Belong?* in order to support particular kinds of argumentation. I want students to notice numbers of sides and angle measures, for example. They tend to do that on the first page of the book. I want students (and teachers!) to develop new language for properties they notice, but don't have the vocabulary for. The spirals (Figure 3.5) provide a reason for everyone to carefully describe previously unnamed properties.

Developing Argumentation Skills

Argumentation is an essential part of mathematical activity. When we teachers ask students to "show your work" or "explain your reasoning," we are really asking them to produce particular kinds of mathematical arguments. When we say we want students to understand why things are true in math, we really want them to be able to argue for (or against) particular ideas or techniques. Young children can produce mathematically viable arguments when they care about the ideas they are arguing for.

FIGURE 3.5
Spirals have properties students can see and describe, but for which they have no standard vocabulary.

Students can become quite invested in these collections of shapes when they know that the ways they see and describe the shapes are valued and useful parts of mathematics instruction. When students are invested, they are more likely to listen carefully to the ways others see these shapes, and to want to defend, extend, and clarify their own thinking. Here is an example of students developing their argumentation skills through *Which One Doesn't Belong?* conversations.

A kindergartner once said that the shape in the lower right in Figure 3.6 didn't belong because, "It looks like a monster truck, and if it flipped over it could still go." This boy was pointing to two unique features of this shape: that it looks like a monster truck (evidently an important characteristic to him), and that it has 180° rotational symmetry. Imagine that you are looking at this truck from the side. It has wheels on top just as it does on the bottom (an unusual property for a truck, of course, but that was the boy's point—if it flipped, it could still go). He didn't have the words for that second property, but it is absolutely the property he was pointing to: this is the only shape on this page that looks the same when you rotate it 180°. One of his classmates raised her hand shortly afterward to present a counterargument: "Monster trucks can't go when they flip upside down." The first kindergartner accepted her correction while defending his vision, "Yeah . . . but this one could!" *Clearly stating a claim*, *producing counterclaims*, and *responding to a critique* are important argumentation moves. *Which One Doesn't Belong?* regularly inspires these

FIGURE 3.6
The monster truck in the lower right can flip over and still go.

moves in a variety of grades because the claims originate with the students. The claims under consideration are meaningful and personal, so students tend to take them seriously.

Exposing Misconceptions

As a teacher, one of the things that sometimes keeps me up at night is the worry that some of my students are producing correct answers without really having any idea what they are doing. I feel that this is one of the biggest disservices I can do to an individual student, so I spend a lot of time thinking about how to bring student misconceptions out into the open. After all, you can only correct your thinking if you're aware that it's not correct, and I can only help you to improve your thinking if I'm aware of your ideas.

Which One Doesn't Belong? is one of several tools I use for exposing misconceptions. The case of the curvy-sided triangle earlier in this chapter is an example of that. While the preceding example imagines a primary classroom, the same set of shapes used in a later-grades classroom where I expect students to already know the definition of a triangle will draw out the misunderstandings present for my students. Similarly, the discussion of diamonds in Chapter 4 is an example of how a *Which One Doesn't Belong?* conversation could draw out misconceptions about the importance of orientation among older students.

HOW TO TALK ABOUT *WHICH ONE DOESN'T BELONG?*

What kind of a classroom resource is *Which One Doesn't Belong?* There is no one answer; it can play many roles. One excellent use is analogous to a read-aloud storybook, one that is long enough to require several days to read in its entirety. In this case, you would spend ten or fifteen minutes each day over the course of a week having conversations about the pages in the order they appear. You would recap yesterday's conversation briefly at the beginning of each session, make predictions, and so on. Most of the ways of talking about a story make sense for talking about this shapes book.

Another use is as one of several classroom routines that cycle over the course of a week. In this case, you might do two or three *Which One Doesn't Belong?* pages every Monday, for example. The first time, you would do the first few pages because I designed them to be a gentle but interesting introduction to the general principles. But after that, you could do whichever pages strike your fancy, in whatever order seems right for you and your students. Likely, too, you'll want to integrate sets that you or your students have made. (The end of this chapter provides advice for getting started.)

Yet another use is as a launching activity for your geometry unit. Do the first two pages, then whichever pages are likely to get students talking about the important themes you'll be studying in the coming days or weeks. Sprinkle in more pages as time allows during the unit (or beyond the unit, in order to keep students' geometry minds active).

Surely you will think of other ways of working *Which One Doesn't Belong?* into your curriculum, and you should feel free to be creative in doing so. I have only two constraints (which of course I cannot enforce): (1) This is not a textbook. While the order is thoughtful, you are not obligated to follow it, nor are you obliged to do all of the pages. There are no prerequisites. (2) This is not a test. *Which One Doesn't Belong?* is great group formative assessment—a way of finding out what your students know collectively. It is not a good summative assessment and should never have grades or other implications for students' future studies. The intellectual risks you ask students to take in a *Which One Doesn't Belong?* conversation have to be supported by their confidence that their answers cannot be held against them.

As with any classroom routine, the exact character of your *Which One Doesn't*

Belong? conversations will depend on your personality, skills, and interests as well as on those of your students, and most of this section so far has assumed that you are working with the whole group. A few explicit words about some ways to structure student-teacher interaction in a variety of classrooms will be helpful in getting you started.

Whole Group

For many teachers, conducting a whole-class conversation in mathematics is a new and intimidating idea. A lot of the interaction that occurs in mathematics teaching follows a familiar pattern: teacher asks a question, student responds, teacher gives feedback. The pattern is so firmly established that we often don't really notice it, nor think to name it. Researchers (e.g., Mehan 1979) have both noticed and named it—the Initiate-Response-Feedback pattern, or IRF (sometimes this is referred to as Initiate-Response-Evaluate, or IRE). *Which One Doesn't Belong?* is a helpful structure for building different, more engaging, and richer patterns of classroom mathematical interactions. Other examples of such structures include Number Talks (Humphreys and Parker 2015) and Problem Strings (Harris 2011).

For a whole-group *Which One Doesn't Belong?* conversation, you want to organize students and the visuals so that everyone can see without straining their eyes or necks. In primary classrooms, this usually means having students on the classroom carpet looking at a projected screen or interactive whiteboard. In upper elementary and secondary classrooms, this may mean rearranging desks so that everyone has a clear view—a semicircle is an especially nice arrangement that allows students to see each other as well as the images.

Some teachers make a habit of writing student observations about each page on top of the images on an interactive whiteboard, creating an annotation of sorts. Others keep track of this information on a different board or on chart paper. Either way, organizing the information in a way that mirrors the structure of the pages—claims about the upper-right shape go in the upper right, and so on—helps everyone to keep track of what has been said, and of questions that have arisen. Simply writing a key word (*square*) or phrase (*all angles the same*), or sketching a quick diagram, or circling key features of the shape are all quick ways to maintain visible reminders of the unfolding conversation for everyone to access.

When I am visiting a classroom, I find that the variation in available classroom media means that I rely much less on writing down the observations we make as a class and much more on recapping the conversation as we go. Whether you are recording student observations in writing or not, summarizing (or having students summarize) the discussion on each page before moving on signals to students that these ideas are important and can help them to retain and use the new ideas. Something such as the following provides a good transition between pages:

> So on this page, Ellie was using the shading of the shapes. Winston talked to us about the direction the shapes were facing, which is the **orientation** of the shapes. Eli described what the shapes are made of, or how they are **composed**. Finally, Molly used the sizes of the shapes' angles to say how they were alike and different. Shading, orientation, composition, and angle measures will be things we'll want to look for on the next few pages, and we'll also want to add to this list of ways we can talk about shapes.

The more you practice this kind of summary, the easier and more natural it becomes. These may even be things you have a lot of experience with in discussing literature or other topics with students. *Which One Doesn't Belong?* conversations give many teachers the opportunity to practice familiar skills in the new context of mathematics.

Small Group

Many teachers have students work through *Which One Doesn't Belong?* sets in pairs or small groups—especially after the routines have been established through whole-class discussions. Typically, this involves students moving at their own pace through a set of prompts the teacher has chosen. If you choose or create about five prompts—and allow enough time for everyone to get through three of them—you'll likely give enough challenge to your faster-moving students while letting your slower-moving students have enough time to engage. Not everyone needs to do all of the same prompts. Depending on your classroom environment and traditions, you may want to have students record some product of their small-group work. This can take place on paper, mini whiteboards, or tablet computers quite naturally. Typing responses on laptops or Chromebooks is likely to impede the kinds of work students will want to do, which often includes drawing diagrams and arrows pointing to parts of shapes.

One-on-One (or Two)

In this case, the *Which One Doesn't Belong?* conversation becomes like reading a book together—a rare luxury for a classroom teacher, to be sure, but such a delight when possible even only for a brief time. You and your conversation partner will want to get comfortable on the floor or beanbag, book in hands exactly as though you were reading a storybook. In a very real sense, you *are* reading a story. In this case you are building your own mathematical narrative together. As you turn the pages, the protagonist shapes develop richer, more complicated and problematic relationships with each other. You and your partner resolve that tension by noticing interesting and surprising properties.

If this kind of opportunity isn't available to you and your students during the school day, consider making time for it after school or at home with your own children, grandchildren, or children of neighbors and friends. While the children in your life outside of school may not clamor often for formal instruction in decimal long division, they (and you) will greatly enjoy talking together about what you notice in these rich sets of shapes. When there is no one right answer, children's mathematical minds are freed to be creative and playful. Seize this opportunity to represent mathematical thinking to the people in your life outside of school.

Classroom Library

Put a copy or two of *Which One Doesn't Belong?* in your classroom library. File it along with the other nonfiction books and allow students to curl up with these shapes to have an internal conversation about geometry properties and relationships.

RESOURCES FOR GOING BEYOND THIS BOOK

No matter what structure you tend to work with, you'll eventually find yourself needing additional *Which One Doesn't Belong?* sets beyond the ones included in this book. There are three useful resources for this:

1. My website, *Talking Math with Your Kids* (talkingmathwithyourkids.com). Type "Which one doesn't belong?" in the search box there, and you will find articles that contain additional prompts I've developed, and that I continue to develop and share for your classroom (and home) use.

2. The *Which One Doesn't Belong?* website (wodb.ca). This website is maintained by Ottawa, Ontario, teacher Mary Bourassa. Teachers (and the occasional student) submit their sets of four objects, and Mary organizes and categorizes them for easy browsing. This website is where a lovely innovation took root—incomplete sets. In this structure, teachers offer three shapes (or graphs, or everyday objects, or whatever) and challenge students to identify a good fourth object for the set. This is a really great way to increase and change the challenge for students. The website wodb.ca is the place to find classroom-tested partial sets.

3. Mary Bourassa also started and maintains a Twitter account: @WODBMath. People (including me) sometimes share their *Which One Doesn't Belong?* sets, together with tales from the classroom, by mentioning this account. Following @WODBMath on Twitter—or just occasionally searching for tweets mentioning it—is a good way to come across new prompts and ideas for your classroom, as well as to engage directly with other teachers playing with this same classroom routine in their classrooms across the K—12 grade spectrum.

4. Also on Twitter, you can share your made sets, your found sets, and your thoughts and questions about *Which One Doesn't Belong?* using the hashtag #wodb.

You'll probably also end up inspired to create your own *Which One Doesn't Belong?* sets for use in your classroom. While this is indeed a challenging task, you needn't be intimidated by it. Consider the successes and struggles you'll have doing this to be powerful mathematical and pedagogical learning. And remember that you're not alone on this journey. Many of us have taken our own versions of it. Up next are tips—based on the very real experiences of teachers in classrooms—for generating your own sets for use in geometry and other topics.

BUILDING YOUR OWN SETS

You certainly could grab four random objects, put them in a 2-by-2 array, and ask your students to say which one doesn't belong. There would be no guarantee that the differences that surfaced in that conversation would be mathematically productive (although they may certainly be productive in other ways, and you could happen upon something interesting purely by chance). But as with most examples you use in teaching, it's usually better to choose them carefully.

Here is a general technique for producing good *Which One Doesn't Belong?* sets:

1. Name four properties, and write them down on a piece of paper.

2. Cover the first property with a pencil or your finger.

3. Identify an object that has the remaining three properties but that does not have the one you've covered up.

4. Repeat by covering the second, then the third, and finally the fourth property.

In this way, you guarantee that each shape has something in common with each of the others, and also a way that it is different from all of them. For example, the shapes in Figure 3.7 came from this list of properties:

- red

- is a square

- is small

- is resting on a side

Careful inspection of Figure 3.7 will reveal that each shape has exactly three of these four properties but not all four.

This is not the only way to produce sets for *Which One Doesn't Belong?* but it provides

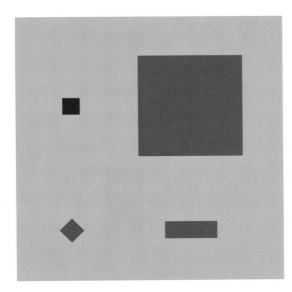

FIGURE 3.7
Each of these shapes has exactly three out of four properties.

a good framework for getting started, and for making sure that there is at least one answer for each of the shapes in your set. I say *at least one* because every time I have built a new set and used it with students, they have told me about interesting properties I didn't notice when I made it. When you ask students to think creatively and freely, they will respond with original and delightful ideas.

I should also warn you of two more things as you set out to produce your own sets.

1. You will probably end up in a situation where your four properties are not possible to combine in the ways you are trying to do. If you find yourself trying to draw "a four-sided polygon with all sides the same length and at least one right angle that is not also a square," you should bear in mind that sometimes these things are impossible. Don't give in too quickly, but don't be too stubborn either. Be ready to revise your set of properties.

2. You may very well end up seeing the world in terms of *Which One Doesn't Belong?* Letters, breakfast cereals, automobiles . . . you might see 2-by-2 arrays everywhere, and find yourself wondering about the similarities and differences among their elements.

It probably goes without saying that students may also begin to see *Which One Doesn't Belong?* sets in the world, and that they may bring these to your attention. Encourage this behavior. Seize the opportunity, and present every student-generated *Which One Doesn't Belong?* set that you are able to make time for.

Last: if you should find yourself in need of a square bagel, shoot me an e-mail. I can tell you where to find one. The evidence is in Figure 3.8.

FIGURE 3.8
Which One Doesn't Belong? **at breakfast**

CHAPTER 4
A Look Inside the Minds of Young Geometers

This chapter explores the overlapping territory of children's informal ideas about geometry and the formal, or "official," mathematics of geometry. To use the language of the van Hiele model, this chapter aims to help you understand, appreciate, and wonder at the richness of kids' level 1 and level 2 thinking. It demonstrates the value of letting children notice new properties of shapes, speculate about relationships, and build informal arguments.

As you read about concavity, spirals, vertices, and diamonds, I hope you will begin to see geometry through children's eyes as well as through the eyes of a mathematician. Mostly, I hope you will come to understand that these two views of geometry are not nearly so distant as the school curriculum might lead us to believe.

In the next few pages, you will encounter definitions of terms that you may not have thought to define before. You will see that definitions themselves are not fixed for all time but change according to need, context, and aesthetics. I hope that your view of mathematics as a field of human inquiry will be a bit richer and more nuanced for working through these ideas. If it is, you can thank the children and teachers with whom I have collaborated in classrooms through conversations about *Which One Doesn't Belong?*

I hope you will read this chapter with a copy of the student book at your side so you can try these ideas out on sets of shapes that don't come up here. Mostly I hope you'll have at the ready at least one other willing and adventurous math mind—a colleague, a friend, a spouse, your offspring, or ideally your students.

CONCAVE AND CONVEX

Students can often identify an important mathematical property long before they can define it precisely. This is the precise opposite of how a lot of instruction in mathematics proceeds. Typically, students are given a definition and asked to apply it. "A triangle is a polygon with three sides. Circle all the triangles on this page." After enough examples of this sequence, students come to believe that this is how math works—math is about correctly applying knowledge other people have generated. This section gives an extended example of how mathematics can develop in classrooms beginning with properties students observe but cannot yet define.

When students work on the shapes in Figure 4.1, one of the most common properties they will cite for the shape in the lower left is that it *doesn't go* in *anywhere.* There are several other ways students phrase this same property. It isn't dented, all of its angles point outward, and so on.

The formal math term for this property is *convex.* In Figure 4.1, the shape in the lower left is the only convex shape. The others are all *concave,* or *non-convex.*

There are two ways to think about convexity. The first is that a shape is convex if each of its interior angles measures less than 180°. A square has four 90° angles, so it is convex. The *sum* of the measures of the angles of any triangle is 180°, so all triangles are convex. The quadrilateral in Figure 4.2 is not convex because the marked angle measures greater than 180°.

FIGURE 4.1
The cupcake page

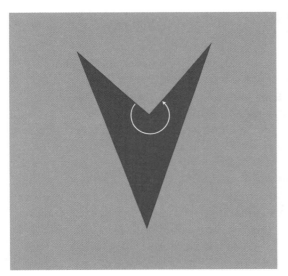

FIGURE 4.2
A non-convex quadrilateral

An angle measuring greater than 180° is called a *reflex angle*. Reflex angles in polygons provide good fodder for conversation in classrooms. Many students will look at the quadrilateral in Figure 4.2 and claim that all of its angles are acute. That's not true, but this common claim can lead to productive mathematics. The claim allows us to notice the need for the term *interior angle*. The following excerpt from a third-grade classroom demonstrates one way to talk with students about this difference.

Tianna: All of the angles of that shape are acute—look at how pointy they are!

Teacher: Let's look at this angle that is kind of in the middle of this shape. Are you talking about the part of it that is filled in bright blue or dark red?

Tianna: Bright blue.

Teacher: OK. Tianna is telling us that this blue part of the angle is acute; it measures less than 90°. And therefore it can't measure more than 180°.

Henry: I saw that too, but I also see the dark red part. It goes almost all the way around, so that must be the part that is greater than 180°.

Teacher: Mathematicians have a word to help us tell these apart. The dark red part is the part that is *inside* the shape, so it is called an *interior angle*. Usually when we talk about the "angles" of a polygon, we mean the interior angles.

When a shape is convex, there is usually no need to specify *interior* angles of a polygon because everyone is paying attention to the same angles. But when a group of people is trying to understand the nature of non-convex polygons, they need this vocabulary in order to ensure everyone is talking about the same thing.

You may expect that the bright blue part of the angle in question would be called an *exterior angle*, and that's a natural thing to expect. But it isn't called that. There is such a thing as an exterior angle, but this isn't it. In fact, that bright blue angle in question doesn't have a standard name. Frequently, I will suggest to students that the rest of our conversations would be easier if we *could* refer to that angle—and others like it in other shapes—by name, and we give it a temporary and idiosyncratic name. Often, we name it after the student who pointed it out to us in the first place, such as a *Tianna angle*.

But what about shapes that are not polygons? Most students will argue that a circle is convex, but the amoeba shape in Figure 4.3 is not. The informal ways of talking about convex and concave apply just fine when the sides are curved: the amoeba looks dented while the circle doesn't, for example. There are no angles to measure in either shape, though, so it's useful to have a second way of defining the term *convex*.

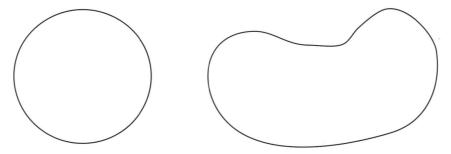

FIGURE 4.3
A circle is convex. An amoeba is concave.

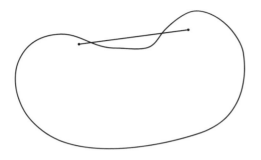

FIGURE 4.4
Drawing a segment between two interior points

A second way to define convex is this: pick any two points inside the shape and connect them with a straight line segment. That line segment either stays completely inside the shape, or part of it crosses outside the shape. If all such line segments stay inside the shape, no matter which points you select, then the shape is convex. If it is possible to pick two points inside the shape and connect them with a line segment that goes outside the shape, as the line segment in Figure 4.4 does, then the shape is non-convex, or concave.

An *exterior angle* of a polygon is an angle that, together with the interior angle, makes a 180° angle at a vertex of the polygon. The angles labeled with letters in Figure 4.5 are all exterior angles. Exterior angles are useful for understanding interior angle sums in polygons. (In Chapter 2, I provided a short proof that the measures of the interior angles of any triangle sum to 180°—the angles labeled A1, B1, and C1 in that proof are exterior angles.)

VERTICES

The shapes in Figure 4.1 are useful for getting students to clearly articulate the meaning of a term—*convex*—that names a property they mostly see and agree on. This same collection can help students better understand a term they likely *don't* agree on: *vertex*.

Frequently, a *Which One Doesn't Belong?* conversation leads to an argument. Not fisticuffs and yelling: a mathematical argument in which students defend conflicting claims

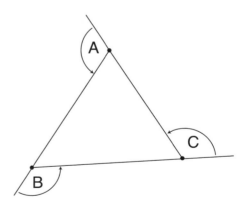

FIGURE 4.5
The labeled angles are exterior angles.

FIGURE 4.6
These shapes spark discussion about the meanings of *corner* and *vertex*.

using logic and evidence. An important role for the teacher in such a situation is to make sure all students understand each of the claims. This serves to keep everyone engaged and learning, and it presses students to state their claims more clearly, usually coming to understand their own ideas better as a result.

The collection of shapes in Figure 4.6 is a great opportunity for bringing out conflicting claims, and for demonstrating the importance of precise vocabulary and clear definitions in mathematical argumentation.

In many classrooms, the shape in the upper right is a launching point for this conversation. A student may say something such as, "That one doesn't belong because it has three corners." The student who says that is noticing the bottom vertex and the two at the top but is missing the vertex that is sort of in the middle of the shape. Often, that student is thinking about the shape as a physical space. If it were a classroom, for example, and you were told to "find a book in the reading corner," there are three choices for where the reading corner might be; no way would that vertex in the middle of the room be considered a "corner" of the classroom.

This is an opportunity to introduce the term *vertex*. One way to do so is by saying something such as this:

Jerome says this shape has three corners, and we talked about how that's sort of true if this were a classroom. But mathematicians think about corners differently.

42

In math, we even have a special term for a corner. We call it a "vertex." If you have more than one vertex, you have two or more "vertices." And what matters for a vertex is that there are two sides coming together, not the exact way that they come together. So in addition to thinking about corners of a room, I also want us to be able to talk about vertices. While this shape would have three corners as a room, it has four **vertices**.

The keys to this message are that (1) it honors the informal ideas children bring to the math classroom while (2) it introduces the more formal ideas and vocabulary. There are many ways to achieve these goals. The preceding is not intended as a script for you to follow with your students. But it would be a shame for Jerome to walk away thinking that his answer doesn't count or is wrong. Instead, I want Jerome (and the like-minded children in all classrooms) to understand that his idea is valuable for helping all of us to notice and discuss an important piece of mathematics. I want Jerome to know that he contributed to the class's knowledge.

After reaching consensus on the number of vertices in the quadrilateral in Figure 4.6, there is still the matter of the number of vertices in the other shapes. In particular, consider the shape in the lower right. In most classrooms, there are two counts that students are willing to defend: zero and eight. This is a good opportunity for a show of hands, so that students can see that others are thinking like them. It is also a good opportunity to ask, *What do you think the others are thinking? If you chose zero, what are the people who chose eight thinking? And if you chose eight, what are the people who chose zero thinking?*

Those defending zero typically argue that a vertex is the place where two straight edges meet. While this shape has some straight edges, they do not meet. Each straight edge meets a curved one, but this is not properly called a vertex.

Those defending eight typically argue that a vertex is a meeting place. It is where two edges come together, regardless of whether those edges are curved or straight.

The important points here are that definitions have consequences and that *people* write definitions. If you define a vertex to be a meeting place of two line segments, then this shape has no vertices. If you define a vertex to be a meeting place of the shape's

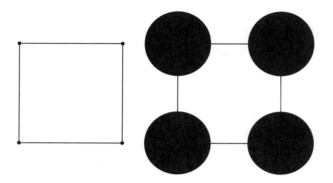

FIGURE 4.7
On the left, four vertices. On the right, four *really big* vertices.

edges, whether those edges are straight or curved, then the shape has eight vertices.

Typical practice in mathematics when presented with this sort of choice is to choose the definition that is least constraining and most useful. There really is nothing bad that happens if you allow curved sides as well as straight sides to meet at vertices, and it makes analyzing a wider variety of shapes easier. For these reasons, mathematicians generally accept that a vertex is a meeting place of edges, whether they are straight or curved.

That said, most of school geometry focuses on polygons, so students will rarely meet vertices that involve curved edges. But rest assured that you will not be doing your students mathematical harm by agreeing as a class that there are eight vertices in this shape, and you won't be doing them harm for future school geometry either. Your state test, for example, will never ask how many vertices are in a shape such as this one. What it will ask—and what is of high importance for the study of geometry—is how many vertices the quadrilateral in Figure 4.6 has. Ironing out the difference between *corner* and *vertex* is important.

In a talk I gave to teachers recently, I presented the options zero and eight to the group and asked my typical follow-up question (which I also recommend you ask): *Are there any other numbers of vertices that we should add to this list?* One teacher raised his hand. I called on him, and he said, "Four." I acknowledged this as a new answer; I hadn't heard anyone yet make an argument for four vertices in this shape. He elaborated by playing with the idea that geometry diagrams such as the one in Figure 4.7 often have dots representing the vertices—"four really big ones!"

MEASURING ANGLES

Whether we call them *vertices* or *corners*, these are the places where we measure the interior angles of shapes.

Most of us who don't grow up to become research mathematicians spend the majority of our geometry studies on polygons. After years of studying polygons with their straight sides coming together to form neat and tidy angles, it may be surprising to hear young children describe the angles of things that do not have straight edges.

The circles that are part of the shape on the left in Figure 4.8 are perpendicular to the square. There is an acute angle at the bottom of the heart (in the middle of Figure 4.8). The cupcake on the right has four obtuse angles. Each of these is a common claim in an elementary *Which One Doesn't Belong?* conversation.

The reason young children make claims about angles in figures with curved sides is the same reason mathematicians do: the vertex at the bottom of the heart looks an awful lot like an acute angle. We shouldn't really treat it any differently just because the sides aren't straight. The spirit of the thing is clear—two edges are coming together; they are going in different directions; the difference in these directions is measurable and should be called an angle. Figure 4.8 shows some angles that are right, some that are obtuse, and one that is acute. Even if you aren't sure how to measure these angles

FIGURE 4.8
Do these shapes have angles? How would they be measured?

precisely, you would probably be able to say which are which.

Mathematicians, however, have the added responsibility of being precise about what it means to measure the angle between two curves, or between a curve and a line. Students cannot quite put it in the language of mathematicians, but they have many of the same ideas. For example, I participated in a version of the following conversation in a third-grade classroom.

While looking at the heart in Figure 4.8, someone objected that you can only measure angles between straight sides.

Iris: But maybe it is straight!

David: No. Those are definitely curved edges. Some hearts have straight sides, but this one doesn't.

Iris: I know. What I mean is that maybe—if you looked really closely—the curves stop curving right before they meet, and they straighten out. So maybe it looks like it's two curves coming together, like if you're looking from here. But if you got really close, it'd be two straight lines where they come together. And then that would be an acute angle.

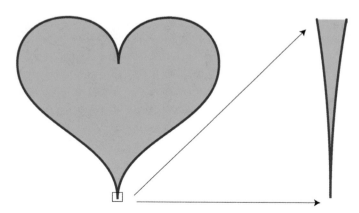

Mathematicians don't speculate about the possibility of the curve straightening out in the way that Iris did. But they do talk about how curves

FIGURE 4.9
The bottom of the heart looks closer to two straight edges when you zoom in.

behave like straight lines when you look very closely. On the left of Figure 4.9 is another heart. On the right is the bottom of this heart zoomed in. The curves look more like straight lines at this scale. Once you have zoomed in far enough that the curves appear to be lines, the angle between those apparent lines is the angle between the curves.

Calculus makes all of this precise by using tangent lines. Children are thinking about the ideas of calculus—and doing very real and very important mathematical work—when they argue about whether curved sides can form angles, and when they wonder how to measure those angles. But don't take my word for it. At www.stenhouse.com/wodb, you'll find a zoomable heart (powered by Desmos): a dynamic version of Figure 4.9 that you can use to explore these ideas yourself. Try zooming in on various parts of the heart and see what happens when you magnify them.

LENGTH OF A SPIRAL

"That one doesn't belong," a first grader told me one day while pointing at the spiral in the upper right in Figure 4.10, "because it's the longest."

Prior to this day, students had talked about this spiral as the *largest* and the *widest*. They had talked about the spiral in the lower left as the *widest in the middle*. But no one had talked to me about the length of a spiral.

I asked what she meant and she said that if you were to straighten out the spirals and compare the results, the one in the upper right would be the longest. This was in a fifteen-minute classroom

FIGURE 4.10
How can we order these spirals by length?

visit, so we didn't take the time to make this comparison in a careful way, but it is certainly worth exploring.

I began to wonder how the lengths of the other three spirals compared, and I have since discussed the matter many times with elementary children and teachers. My experience is that students do not immediately see that the upper left and lower right spirals are the same length as each other, and it is worth letting students wonder about and explore this question.

The spiral in the lower left is a particularly challenging case. Students tend to quickly agree that it is shorter than the one in the upper right, but comparing to the one in the lower right is more challenging. Here are two arguments, one for each comparison:

The spiral in the lower left is longer than the one in the lower right. If you think about the spiral as going from the inside out, this one starts out longer because it has a wider inside. The end of the spiral is in the same place as for the lower-right spiral, so their ends are the same length. The bigger beginning means the lower-left spiral is longer.

The spiral in the lower right is longer than the one in the lower left. If you think about the spiral as going from the outside in, they start out the same. But the one in the lower right has to "reach" farther in toward the center. Think about how you need longer arms to reach something farther away. In this case, the inner end of the lower-right spiral is farther away, and so the spiral must be longer to get that far.

Over the course of about two weeks and a half-dozen classroom conversations about these spirals one spring, I heard students put forth various forms of these arguments many times. I remained frustrated by our inability to answer this question definitively.

I could devise two ways to make the comparison—the kindergarten way and the calculus way. In the kindergarten way, you take a piece of string and wind it around the spiral, then straighten it out. (Actually it turns out that pipe cleaners are much better tools than string for this task!) The calculus way is to put the spiral on the coordinate plane, write an equation for the spiral, and then use general techniques for approximating the length of a curved path. Before reading further, you really should take a few moments to measure these spirals yourself (in the kindergarten way). Which one really is longer, and by how much? Is the difference you find by these approximate measurement techniques big enough that you're sure it's correct (and not just an error of precision)? Is the difference big enough that kindergartners could be sure of it?

Each of these techniques—direct measurement and calculus—can settle the question of which spiral is longer. I have found that there are relatively few questions in mathematics that can be answered with kindergarten ideas and also with calculus ideas, but not with anything in between. The lack of a middle school–level argument for this comparison was frustrating for me. It was resolved one day by a college math student of mine.

Before I tell you what my student said, take some time to consider the question yourself. In addition to the value in answering the question as posed, I also want to point out the significance of the kind of question that it is, and I want to point out its origins. *How can we know in a more rigorous way than physically measuring, but not requiring sophisticated mathematical tools, which of two spirals is longer?* This is a well-posed and wholly original mathematical question—one that has likely never been asked in precisely this way, and whose origins are in the mathematical activity of schoolchildren. It is a question that (until now) could not be answered by looking it up in a book or searching the Internet.

Here is my student's argument: Each of these spirals goes around two and a half times. Think of that as five half-turns, as in Figure 4.11. Each half-turn is close to, but not quite, a half-circle ("not quite" because the radius is constantly changing, and a circle has a constant radius). If you replace each not-quite-half-circle with an actual half-circle, you won't change the length of the spiral by very much. But each half-circle for the lower-left spiral will be a bit bigger than each half-circle for the lower-right spiral. Because all five half-circles on the left have a bigger radius than their corresponding ones on the right, the spiral on the left is longer.

This is a sophisticated argument, but it uses only the ideas of upper elementary and middle school. It is a more precise argument than either of the two initial ones (about

FIGURE 4.11
Two spirals, each made of five half-turns

"wider insides" and "reaching farther in"). Furthermore, this argument can be made even more precise. My student built the argument on half-circles, but you could say the same about quarter-circles, or any smaller piece of a circle. In fact, the calculus solution boils down to taking many very tiny parts of a spiral, treating each as part of a circle (each with a slightly different radius), measuring each part, and then adding these measurements together. The calculus technique is essentially a refined version of the middle school technique.

And, in fact, whether you measure with the techniques of kindergarten, middle school, or calculus, the lower-left spiral is a bit longer than the one in the lower right.

COMPOSING SHAPES

Young children are attuned to what shapes are made of. The two shapes in Figure 4.12 are especially likely to get students talking about composing and decomposing shapes. The shape on the left is typically a semicircle and a trapezoid in the eyes of an elementary school student (or a circle and "another shape," depending on the vocabulary students have ready at hand). The one on the right more commonly elicits debate.

In most classrooms, someone will venture the claim that the shape on the right is made of a square and four circles. After a thoughtful pause, another student will often challenge this claim. This student may point out that if you removed four circles from the shape, you would not have a square left behind, as on the right in Figure 4.13.

FIGURE 4.12
Two shapes that children see as being built from other shapes

Alternatively, a student may notice that this shape is built with a square, but that it is not four complete circles that are added on to complete the shape; it is four circles with missing wedges, as on the left in Figure 4.13. Possibly the first student will defend the original claim by saying that the circles are glued on top of the square, or vice versa.

This kind of conversation is helpful for developing students' spatial skills and for changing perspectives to allow students to consider alternative viewpoints. You might even ask whether anybody sees *other* ways of composing this shape after two or three ways are on the table.

Composing and decomposing shapes is an especially important milestone for kindergartners and first graders, and it represents a step up in sophistication from

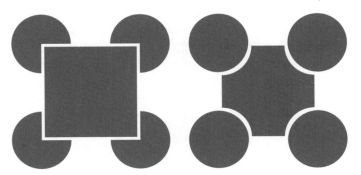

FIGURE 4.13
Two different ways to decompose a shape. Neither one consists of a square and four circles.

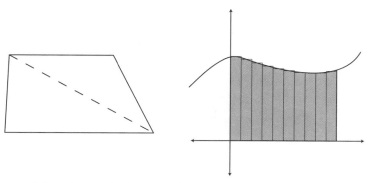

FIGURE 4.14
A middle school and a calculus diagram showing the utility of decomposing shapes

recognizing shapes in the world. Learning to see squares on the street, in the classroom, and in the cracker aisle of the grocery store is one thing. Learning to see squares as building blocks for other geometric figures is more challenging.

Young children need a variety of experiences putting shapes together and taking them apart. Tangrams and pattern blocks are two examples of common school experiences children have with these skills. Minecraft offers many students rich experiences of composing shapes in three dimensions.

In later mathematics courses, composing and decomposing shapes plays an important role in measurement. The left-hand side of Figure 4.14 shows a typical middle school diagram in which the formula for the area of a trapezoid depends on decomposing a trapezoid into two triangles. In fact, most work with area in the later grades requires decomposing shapes (including circles and even ellipses) into triangles and rectangles. On the right-hand side of Figure 4.14 is a typical diagram from a calculus textbook that shows a complicated region being approximated by a decomposition into rectangles.

Young children are naturally curious about composing and decomposing shapes. Building on that curiosity by giving students opportunities to play with—and to discuss— the ways shapes are put together and can be taken apart pays dividends throughout their mathematical careers.

"COULD BE A SQUARE"

"That one doesn't belong," the first grader said, pointing at the rectangle in the lower right in Figure 4.15, "because all of the others on this page could be a square, and that one can't."

The boy said that the others *could be* squares. At that moment, what I heard him say was that the others *are* squares. Several minutes later another student said something similar about a square oriented at an oblique angle on a different page. "That one could be a square."

This second time I heard the words more clearly. I paused and told the students we were going back to an earlier page because I just realized something important. We returned to the page in Figure 4.15, and I said to the student who had originally voiced the idea, "You said that the shape in the lower right didn't belong because all the others could be squares, but this one can't be. Is that right?"

He agreed.

"OK. So what I need to know is this, and we'll do it by a show of hands, first graders.

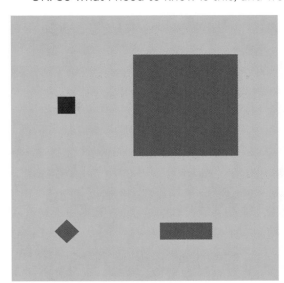

I need to know whether you think this shape in the lower left *is* a square as it is now—that's one option—or whether it *could be* a square if we turned it, but it is not now a square."

I went over the options again, made sure everyone understood what I was asking, and then had them raise their hands.

I have since taken this poll many times in kindergarten through second-grade classrooms. Each time, nearly all of the students raise their hands for the

FIGURE 4.15
"All of the others on this page could be a square, and that one can't."

potential of that shape to be a square, but not for it actually being a square.

I have come to understand that talking about this difference is more important than defining it away. As a teacher, it can be tempting to lament the knowledge our students don't have, and it can be difficult to celebrate the incomplete and messy process that builds this knowledge. For a primary grades student to notice that the shape stays the same is an important step to knowing that the *name* of the shape also stays the same.

It is an important fact that everyone in this classroom agreed that the non-square rectangle in the lower right was *different* from all of the others in that it couldn't be a square. There are several ways to transform that shape into a square. You could squish it from side to side. You could stretch it vertically. You could cut off the ends. Young children have plenty of experience with squishing, stretching, and cutting. By acknowledging that the shape in the lower left of Figure 4.15 *could be* a square but that the shape in the lower right cannot be, these children show that they understand that rotation is a different kind of transformation from squishing, stretching, and cutting. They do not yet understand that rotation doesn't change the way we classify shapes in geometry, while squishing, stretching, and cutting each may change a shape's classification.

But dig just a bit deeper into children's minds. They are sorting out the many different scenarios where orientation matters, and where it does not. When they are solving puzzles with pattern blocks (as in the task in Figure 4.16), orientation matters. When

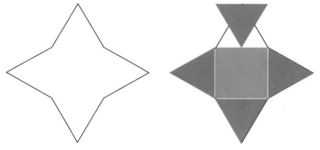

FIGURE 4.16
Orientation matters when solving puzzles with pattern blocks.

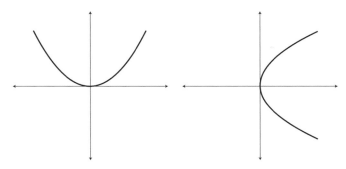

FIGURE 4.17
Orientation matters in algebra—the graph on the left is a function; the graph on the right is not.

they are writing the digits 5 and 2, orientation matters. Even much later on, in algebra, orientation matters. The graph on the left-hand side of Figure 4.17 is a function, while the graph on the right is not; they are the same shape but oriented differently.

When we classify shapes in geometry, orientation doesn't matter. A square is still a square, no matter its orientation. But why is this? It is because geometry defines shapes by the relationships of their parts to each other, not their relationship to space. A square is a four-sided figure with all right angles and all sides congruent. There is nothing in its definition that refers to the space in which the square is situated, nor to the objects around it.

Puzzles, by contrast, are all about spatial relations. A puzzle piece only goes into place when it is in the right position and its parts line up with those of the adjacent pieces. Digits, like letters, have required positions and orientations. The graph of a function depends on points that are positioned along a horizontal (*x*) and vertical (*y*) axis.

Rather than gloss over the question of the difference between "diamonds" and squares, we can use students' ideas about these differences to examine the important question, *When does orientation matter?* Children's ideas about answers to this question will evolve over time. As teachers, we can have greater impact on the direction of that evolution by making those ideas public, and by taking the time to examine them and even to argue about them.

A Few Words About Diamonds

My thinking about the term *diamond* in geometry has evolved, and this evolution is entirely due to the thinking of children I've worked with.

In everyday English, when people describe shapes as diamonds, they typically mean something convex that has between four and eight sides, a vertical line of symmetry, and is oriented with a vertex at the bottom. The shapes in Figure 4.18 are all examples of diamonds. You play baseball on a diamond.

I used to think that letting students use *diamond* to describe shapes in geometry was letting them get away with something less rigorous, or possibly letting misconceptions slide. Now I know that students are still learning when to attend to orientation and when not to. Now I know that *diamond* is a term that has a shared meaning for students—a meaning that comes from their experiences with shapes outside the classroom. While a diamond isn't a well-defined object in geometry, I know that we need to help students along to this understanding, and that conversations such as those sparked by a well-

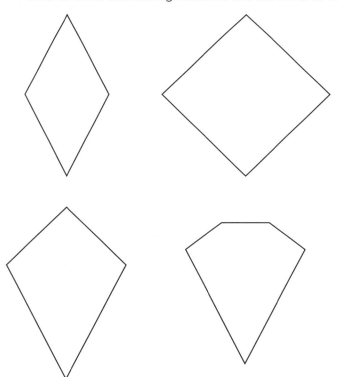

FIGURE 4.18
Some diamonds

designed *Which One Doesn't Belong?* set are tools for exposing the differences between everyday English and formal mathematics.

Now I know that informal mathematical terms coming from students' outside-the-classroom, lived experience are just diamonds in the rough.

WHAT COUNTS AS A PROPERTY?

People often characterize mathematics as being absolute, certain, and unchanging. There is a kernel of truth to this, of course. Two plus two is four and all that. But math takes place in cultural contexts, and it has different fields that pay attention to different properties. The properties and relationships that are valued in these different contexts and fields shift in ways that may seem subtle and surprising to someone who expects mathematical truth to be eternal and fixed. This section aims to make these claims less abstract, and to help you and your students appreciate that mathematicians deal with aesthetic concerns, just as poets and other artists do. In all areas of life, people do things because they are beautiful, elegant, or just *feel* right. Math is no different. To illustrate, we'll consider several examples from *Which One Doesn't Belong?*

In particular, we'll look at a handful of properties. It will feel satisfying to call some of these "properties," while it will feel odd to apply that name to others. This short exercise will help to clarify what it means to talk about aesthetics in mathematics.

> **Example A.** The set of shapes in Figure 4.19 often elicits a claim that the shape in the upper left doesn't belong because it's the only triangle; the other shapes on the page are all quadrilaterals. In this case, the property *is a triangle* contrasts with the property *is a quadrilateral*.

FIGURE 4.19
Three quadrilaterals and a triangle

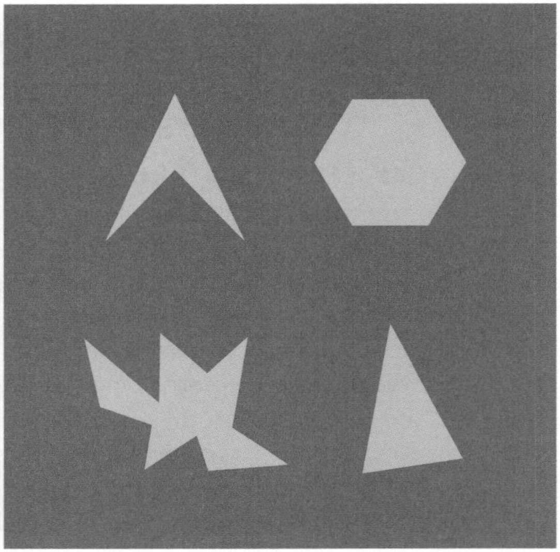

FIGURE 4.20
One quadrilateral and three non-quadrilaterals

Example B. The set of shapes in Figure 4.20 will sometimes elicit a claim that the shape in the upper left doesn't belong because it's the only quadrilateral. The other shapes on the page are not. In this case, the property *is a quadrilateral* contrasts with the property *is not a quadrilateral*.

Example C. The set of shapes in Figure 4.21 usually elicits a claim that the shape in the lower left doesn't belong because it is convex. The other shapes on the page are not. In this case, the property *is convex* contrasts with the property *is concave*.

Frequently, when working with adults (and occasionally with children), someone will say that something feels wrong about Example B. How can *is not a quadrilateral* be a

FIGURE 4.21
One convex shape and three concave shapes

property that unites a diverse group of shapes? Don't properties have to characterize qualities shapes have, not qualities they don't have?

This is a reasonable objection. And contrasted against Example A, the difference is clear. The triangle has a property with a name—it is a triangle. Similarly, the other three shapes have something satisfying in common—they are quadrilaterals. *Three sides* contrasts clearly with *four sides*.

The contrast between Examples B and C is less stark, however. Another term for *concave* is *non-convex*. Concave is a property defined in terms of what a shape is not. There are many such properties in mathematics. Odd numbers are not even, for example (and my daughter, at age eight, refers to *even* and *uneven* numbers—a perfectly mathematically correct thing to do!). Any time you sort things into two categories that don't overlap (what is called a *binary sort*), you are sorting according to *has this property* and *doesn't have this property*. Sometimes these two categories each have a unique name—such as even and odd or concave and convex—but not always.

The truth in the objection to Example B lies in aesthetics, not logic. Sorting shapes according to quadrilateral/not quadrilateral, and then talking about *not a quadrilateral* as a property of a shape, is logically no different from sorting shapes according to convex/concave, and then talking about *concave* as a property on its own terms. That these two cases seem different, and that prevailing mathematical practice tends to value one of these sorts as more important than the other, points to the role of aesthetics in mathematics.

When mathematicians do their work—whether they are academic research mathematicians collaborating on a paper to be published, or a group of kindergartners discussing a set of geometric shapes—they make aesthetic and practical decisions about what is going to count in their investigation; they make decisions about what matters and what does not. These decisions are social, and they are open to revision later on. They are not absolute nor certain nor unchanging.

Indeed, one of the social decisions that was made in school geometry was to value and pay attention mainly to the geometry of polygons (in two dimensions, and to polyhedra in three dimensions). *Which One Doesn't Belong?* encourages teachers and students to push back on this consensus. This book is in part an attempt to free children's minds in school geometry and to allow them to explore the wonder of many other important ways of structuring two-dimensional space. It is designed to give you and your students explicit permission to talk about hearts, diamonds, cupcakes, and spirals as mathematical objects.

CHAPTER 5
Answers Key

You are probably accustomed to math books having an answer key. Well, this one has an *answers* key because it is where I share with you some of the answers that students create for why each shape doesn't belong. It is my sincere hope that this chapter does the opposite of what answer keys are normally designed to do. I hope this chapter will serve to open up possibilities for you and your students, rather than close them off. I also hope this chapter will help you know what kinds of answers to expect your students will provide for each page of the student book. Even more, though, I hope your students will surprise you with answers I haven't anticipated here. For far too many people, math has been an exercise in getting the answer in the back of the book. *Which One Doesn't Belong?* is about breaking that mold.

Instead of telling you what the right answer is, I am using this answers key as a place to share the conversations I have had with children and teachers. Instead of telling you what to think in these pages, I am sharing with you how others have thought.

I have two main purposes here:

1. To provide you and your students with insights and ideas to discuss that may not arise in your classroom today. I want the answers key to expand the ways you see and think about these pages.

2. To help you and your students get unstuck. Some of the collections on these pages are challenging. It is fun to leave a question unresolved for a day or two, but there may come a time when productive struggle becomes distracting frustration. The

answers key is here to provide relief from that frustration. Think of it as bringing more students into your classroom to help you think in new ways. This answers key provides the collective wisdom of dozens of classrooms and hundreds of students.

So here is the answers key. I hope you'll add to it by writing in the margins the things you and your students notice, and that you will take ownership of it. I hope that your version of this answers key will be richer than mine and show evidence of learning, creativity, and mathematical insight.

This opening page is designed to be open to all levels of background knowledge. In the classroom, it is important to make sure students understand that the goal is to notice and discuss difference and sameness, rather than to use the fanciest math language. The examples in the opening pages of the student text model this, and classroom conversations ought to as well.

This is the only triangle. Primary grade students will often say it is the only one that has three corners, and they tend to notice corners rather than sides. In any case, it is generally not obvious to the youngest students that the number of sides and the number of vertices (or corners) are the same.

A common response to this shape is that it is the only one that cannot be made from the triangle in the upper left. Two such triangles make each of the lower two shapes, and the triangle in the upper left is made of itself. Students will often say that this is the only square. Older students may say it is the only one with right angles.

The most common response to this shape is that it is the only one that is not colored in. Some students reject this claim and argue that it is colored in—it's just colored the same as the background.

For many children, the primary distinguishing feature of this shape is orientation. They have several ways of talking about that. Young children will describe this shape as being the only one that isn't pointing up or down but is pointing off to the side. Often, students describe the bottom of this shape as flat, while the bottoms of the others are pointy. Some students will say that this shape seems to be leaning to the side. Others will say that it is the only one resting on a side—the others are standing on their corners.

This page aims to focus student attention on geometric properties, such as the meaning of square, while maintaining less formal entry points (such as size, color, and orientation) so that all students can participate.

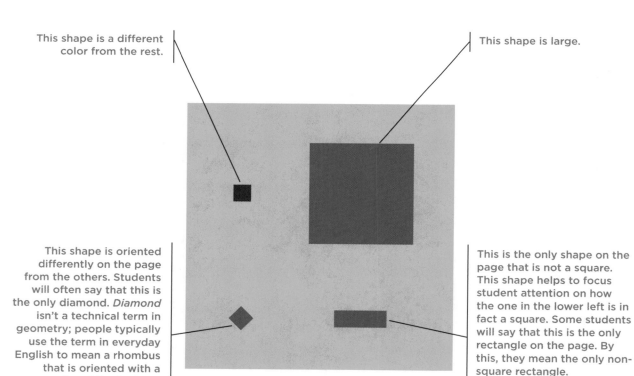

This shape is a different color from the rest.

This shape is large.

This shape is oriented differently on the page from the others. Students will often say that this is the only diamond. *Diamond* isn't a technical term in geometry; people typically use the term in everyday English to mean a rhombus that is oriented with a vertex at the bottom.

This is the only shape on the page that is not a square. This shape helps to focus student attention on how the one in the lower left is in fact a square. Some students will say that this is the only rectangle on the page. By this, they mean the only non-square rectangle.

Students may say that this is the only shape that can be built from three squares.

This page is a good one for discussing the meaning of the term *vertex*. See Chapter 4 for detailed discussion of the relationship between vertex and corner, and for an extended discussion of the meaning of vertex in mathematics.

This is the only shape with no straight sides. Students often say that it's the only shape for which they have a name (a "heart").

This is the only shape with no curved sides. It is the only polygon and the only quadrilateral.

This is the only shape that is convex. The opposite of convex is concave, so the other shapes on this page are concave. Elementary students don't need to use these terms, but they will notice these properties on this page even if they can't express them formally. For example, students might say that this shape is the only one that—if it rained— wouldn't collect water on the roof.

This is the only shape that looks like a cupcake.

Students enjoy this shape a great deal. They frequently have very creative conversations about it. Some sample ideas from students: It is the only shape that is made from four circles and a square. It is the only shape that doesn't have any corners. It is the only shape with more than one line of symmetry. It looks like a monster truck, and if the monster truck flipped it could still go.

This page was designed to bring out conversations about quadrilaterals in general and squares in particular.

Students may say that this is the only rectangle. It isn't, though. The lower left shape is also a rectangle. It is correct to say that this is the only non-square rectangle, or that it is the only oblong rectangle. It is also the only figure that has vertical short sides.

This is the only shape with curved sides, so it is the only one that is not a polygon. It is the only concave shape on this page and the only one with acute angles. (In Chapter 4, I discuss what it means to think about angles between curves.) This is the shape with the sharpest points.

This is the only square. Students may notice that you could obtain each of the other shapes by cutting a square with scissors. In that case, this is the only square that hasn't been cut.

Students may say that this is the only shape that doesn't have four sides. It is the only octagon and the only shape with obtuse angles. Students may say that this is the only shape you could make by cutting off the corners of a square.

This is a page about triangles. The shapes are placed on the page at arbitrary angles in order to reduce the amount of attention students are likely to pay to orientation (contrast with page 5, where orientation is likely to be an attribute students use to distinguish the shape in the lower right).

This is the only triangle on the page. Each of the other shapes has some—but not all—of the defining properties of a triangle. The shape in the upper right has the wrong number of sides to be a triangle. The one in the lower left has three sides, but it is not closed. The shape in the lower right has three sides and is closed, but the sides are not straight.

This is the only shape with more than three sides. It is the only quadrilateral and the only rectangle. This is the only shape with all right angles.

This is the only shape that is not closed. Students may talk about this shape as an incomplete triangle, or more rarely as an incomplete quadrilateral.

This is the only shape that has curved sides. Students sometimes refer to this shape as a triangle.

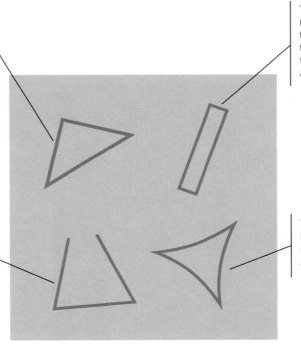

An important feature of conversations focused on this page is that no one has an unfair advantage due to prior exposure. Opportunities to learn vocabulary about quadrilaterals may vary widely even among students in the same class, and certainly mastery of that vocabulary varies greatly. But no elementary student has studied vocabulary for properties of spirals. They can all notice properties of spirals, but none have studied them. Everyone is on a level playing field.

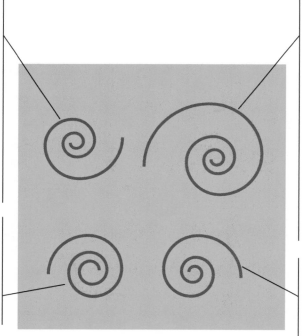

This is the only one that opens upward; if it rained, water would collect in this spiral but not in the others. Students may say that this is the only snail on its back, or that it is an upside-down spiral. Students may notice that this spiral can be reflected over a horizontal line to perfectly match the lower right one. If this comes up in class, other students may need to investigate whether it is true.

This spiral has the largest interior width. The rate at which this spiral shrinks (or grows, depending on how you view it; see the discussion of the lower right spiral) is different in this spiral than the others.

Measuring horizontally, this is the widest spiral. This spiral completes three full revolutions; students often say that it goes around more times than the others. If you look at the directions the two ends are pointing, this is the only shape that has them pointing in opposite directions. The others have both ends pointing up or both pointing down.

Students sometimes think about the lengths of the line segments that would result from straightening the spirals. In that case, this one is the "longest" spiral.

This spiral turns in the opposite direction from the other three.

A fun conversation to be had in classrooms is whether students think about the direction of a spiral being from the inside out or the outside in. Most classrooms are nearly evenly split on this question, and there is no right answer. Going from the inside out, this is a clockwise spiral while the others are counterclockwise. Going from the outside in, these are reversed.

This is a page about quadrilaterals. As with the "triangles" on page 23, these shapes are on the page at arbitrary angles in order to decrease the likelihood of orientation being a major feature that students attend to.

This is the only shape with six sides (i.e., the only hexagon). It is the only concave shape. Students may say this is the only shape that looks like a letter (and they may debate whether that letter is *V* or *L*). It is the only one that looks like a broken or folded rectangle.

This is the only rectangle. It is the only shape with all right angles. It is the only shape with two lines of symmetry.

This is the only shape with no right angles. It is also the only shape with no lines of symmetry. Like the rectangle in the upper right, it does have rotational symmetry. It is the only shape with two obtuse angles. (The shape in the lower right has one.)

This is the only shape with no parallel lines. Students may notice that it is the only shape with a vertical line of symmetry. It is the only kite.

There is really no way for students to know this, but it is an interesting fact: this shape was made by dissecting an equilateral triangle into three congruent parts. That means this is the only shape on this page that can be copied and then built into a triangle.

This page is surprisingly challenging. The shapes are mostly familiar and friendly, but this means that they have many things in common. In turn, these commonalities make the differences challenging to tease out and to state precisely.

In many classrooms, this is the last shape that students are able to distinguish from the rest. This shape has the most lines of symmetry and is the only one with a horizontal line of symmetry. It is the only convex shape on this page that tessellates. It is the only shape that bees make and the only one that is included in the familiar sets of pattern blocks common to American elementary classrooms.

This is the only shape with right angles. It is the only shape that does not have all of its angles the same size. It is the only concave shape. Students may say that it looks like it was made from a big square that has been cut up, or alternatively that it was built from three smaller squares.

This is the only shape that has no two sides the same length (it is scalene). Students may notice that it appears to be leaning to the left, which the others do not. This is the only shape without a line of symmetry.

This shape, a pentagon, has a different number of sides (or vertices) from the others. It is the only one with no parallel sides. Students may say that this is the only one that looks like a house or that is shaped like a famous government building.

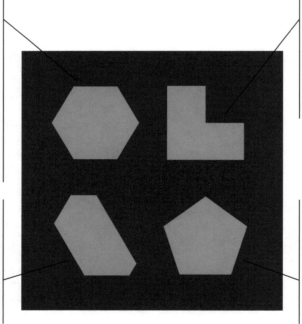

This is a challenging page that is worth revisiting over time. In particular, stating properties that distinguish the two shapes on the left from each other, while also naming their commonalities with the shapes on the right, is difficult.

This is the only quadrilateral. In some classrooms, this observation leads to useful discussion about what property the other shapes have in common. If this one doesn't belong because it is the only quadrilateral, does that mean that the other shapes share the property of not being quadrilaterals? Does that count as a property? Should it?

This is the only shape with all obtuse angles. It is the only shape with all sides the same length and with all angles the same measure, so it is the only regular polygon.

Students will often say that this is the only "crazy" shape, as nothing like it tends to appear in the usual study of school geometry. This is the only shape for which most students don't know a specific name. It is the only dodecagon. It is the only shape that has no lines of symmetry.

This is the only triangle. It is the only shape with an odd number of sides. It is the only one with all acute angles.

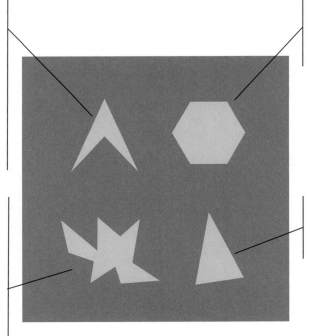

This is a page about polygons. Its placement at the end of the book is intentional, but of course you may use it at any time. The reason it comes at the end is that—for most people—a general and abstract category such as polygon is the end product of encountering many specific examples. Children have many experiences with squares, rectangles, and triangles, both in and out of school. The mathematical knowledge people have from their everyday lives is typically informal—people typically don't have a precise definition of the shape they call a diamond. The discipline of mathematics relies on formalizing these ideas. This means stating precise definitions and using logical relations to verify truth. The early parts of the book are structured to allow students to formalize objects they know something about—to refine the definition of a square, for example, and relate squares to rectangles—before trying to understand the larger category (polygons) these familiar objects belong to.

Even students who don't know the word *polygon* may recognize the shape in the lower right as being the only example of the kinds of shapes they typically see in math class and math books. They may say it's the only one that is "an actual shape" and that the others are just "designs."

Students may say that this is the only shape that crosses itself, or that it is the only one that looks like two shapes glued together. The formal term is that this is the only *nonsimple* shape. (*Simple* curves don't intersect themselves.)

Students may argue about the number of sides this shape has. Does it have four sides, two of which intersect? Or does it have six sides because it is built of two triangles? This can be a distinguishing feature: this shape is the only one on this page whose number of sides is arguable.

This is the only shape that isn't closed. Students may say this is the only one that looks like a paperclip.

This is the only shape with curved sides instead of straight ones. It is the only shape with symmetry.

This is the only pentagon and also the only polygon. Students may say it's the only shape, or the only normal shape. As noted earlier, they may recognize this as the only shape that belongs in math class even if they cannot yet say precisely why.

This page plays with the idea of *one*. Now that you and your students have considered a bunch of ways that individual shapes can be alike and different, it is fun to think about ways collections of shapes can be alike and different.

This is the only page that has all of its shapes colored in. It is the only page with three equiangular shapes.

This is the only page that has both colored and uncolored shapes. It is the only page with more than one quadrilateral. It is the only page with shapes carefully balanced on their points. It is the only page with a square. This is the only page that consists entirely of shapes that tessellate.

This is the only page that has no polygons, and none of whose shapes can be colored in. It is the only page whose shapes are all curvy.

This is the only page with more than one closed, unshaded shape. It is the only page with more than one shape that has right angles. It's the only page with all shapes sitting at unusual angles. It is the only page with both curved and straight-sided shapes.

References

Harris, Pamela Weber. 2011. *Building Powerful Numeracy for Middle and High School Students.* Portsmouth, NH: Heinemann.

Humphreys, Cathy, and Ruth Parker. 2015. *Making Number Talks Matter: Developing Mathematical Practices and Deepening Understanding, Grades 4–10.* Portland, ME: Stenhouse.

Lakatos, Imre. 1976. *Proofs and Refutations: The Logic of Mathematical Discovery*. New York: Cambridge University Press.

Mehan, Hugh. 1979. *Learning Lessons.* Cambridge, MA: Harvard University Press.

National Council of Teachers of Mathematics (NCTM). 1989. *Curriculum and Evaluation Standards for School Mathematics*. Reston, VA: NCTM.

Van Hiele, Pierre M. 1985. "The Child's Thought and Geometry." In *English Translation of Selected Writings of Dina van Hiele-Geldof and Pierre M. van Hiele,* ed. David Fuys. New York: Brooklyn College School of Education.